THE AMERICAN SCIENTIST AND DIPLOMACY

DR. MARIO NABLIBA

978-1-965552-43-8 (Paperback)

BOOKWRIGHTS
HOUSE

admin@bookwrightshouse.com
☎ (213) 286 6700

CONTENTS

CHAPTER 1

American Institute IDPF

IDPF MEANS INTEGRITY DATA Protection Forensic which refers to my first book.

Understanding of science is the joy of our life endeavors. This phrase leads me to quote with my humble idea:

> How many times a day I remind myself that my life depends on the labors of other men, great inventors, scientists living and dead, for the betterment of this nation, and that I must exert myself in order to give, in the measure as I have received, so much on American soil, and I am still receiving …**Mario Nabliba as reflected by Albert Einstein into this New Trend book.**

CHAPTER 2

Tracking and Managing Our Health

THIS LEADS RIGHT INTO tracking or managing the most threatening situations of healthcare, i.e. heart disease. This information will give our scientific community a reality wakeup. Indeed, we must strike against cardiovascular disease as the number one killer in our nation.

Numbers:

The average heart beats 60-100 beats per minute at rest.

Therefore, in a year it would be 31,536,000–52,560,000 just at rest. That amount would be increased depending on the amount of exercise that person does.

1 million	Amount of blood pumped by the heart in an average lifetime, in barrels
35 million	Number of times the human heart beats in one year
60-100 beats per minute at rest.	The average heart beats

2.5 billion	Approximate number of times the heart of a 70-year-old has beaten
60,000	Length of all blood vessels stretched end to end, in miles
128 million in 2020	Number of Americans with a Cardiovascular Disease, including stroke
48.1%	high blood pressure (approximately 20% of the population unaware)
Close to one million in 2022	Number of people dying annually of cardiovascular disease, including stroke, in the United States (approximately 35% of all deaths)
19.8 million in 2021	Number of people dying annually of cardiovascular disease, including stroke, around the world
$320 billion	Annually economic cost of cardiovascular disease, including stroke in the United States (Including direct medical costs and lost productivity)

I learned during my research that "among the numerous dietary ingredients to which our genes are poorly adapted are processed foods and refined sugars ($C_6H_2O_6$), eating substantial amounts of saturated fats, found mainly in processed foods and such animal foods as cream, whole milk and fatty meats. If you are not Gene APOA1, which engages in regulating the body production of HDL cholesterol, as known in more than 30% of people, you raise the blood level of cholesterol. All may know that excess cholesterol in the body deposits in arteries, including the coronary arteries that lead to the heart. (It Is Not Just Your Genes book at the Center of Excellence for Nutritional Genomics at the University of California, Davis). [Ruth Debusk, PhD, RD and Yael Joffe, RD(SA),2006, pp.26-28.]

CHAPTER 3

Washington DC Research

How ONE CAN BE able to improve is through use of efficient technology by using diplomacy.

As an American scientist, my book really has as its main goal the use of diplomacy as the most efficient practice.

To embrace the use of information technology, technology is exactly the excitement of a new generation so it can be a part of our new culture, meaning who has seen this kind of skill set in our background in IT which will be an easy approach to understand, talk or to agree upon of diplomacy policy as well

Example: Coca-Cola-coke-freestyle-soft-drinks-beverages-

An understanding of technology these days is a luxury to be able to know how to use devices and explore the most easily scientific invention gift.

I was in Washington DC, and I went to a restaurant for dinner. After a few minutes of being seated, this young kid walked into the restaurant with a flavor cartridge in hand to change the new model of the Coca-Cola's Machine, for soft-drinks-beverages, called: COCA-COLA-FREESTYLE. Machines can contain as many as 146 flavors dispensed. This kid's name—Brian Davison, is 12 years old according to his information and admired by all the other people that were not familiar with this new electronic–COCA-COLA-FREESTYLE. One customer walked over to him and asked about how the machine worked. I am very familiar with this technology because of being a scientist, doing research I learned from an article in USA TODAY; please see the video:

*http://www.usatoday.com/story/money/business/2014/04/14/
coca-cola-coke-freestyle-soft-drinks-beverages/7478341/*

which is a great innovation from COCA-COLA.

Because I was drafting some of the work of this book, I was inviting him to be part of this book. He was a bit reluctant to talk with me because I was writing a book, and I decided to leave him alone. This little history is a must for this book, because it is proof of how many of this young generation really are in this new century. Like being born with a double helix (DNA), type "Information Technology's GENES" (IT background). Because of this, the young man, Brian, had been there handling the machine with nearly every customer admiring him as he was changing cartridges to the Coca-Cola touch screen system which serves a variety of flavor via one touch on the screen.

After Mike, the dad, walked in, I asked him if he would let me talk with Brian and he said of course and then encouraged him to talk with me. Mike explained the technical literacy of his son, which has amazed us. This is an example of diplomacy and culture and the attraction of information technology (IT) that we are all visioning in this new century, which must be embraced by Information Technology Diplomacy.

In fact, our culture attracts people's attention when talking with someone who possesses an understanding of information technology.

Brian Davidson is a smart young kid!

How one can be able to improve the technology by using diplomacy: American scientists have one of its main goals to use diplomacy, to be the most efficient practice, and this also encourages the embracement of the use of information technology. Technology is exciting for new generations so it can be a part of our normal skill set; our background in IT makes it easy to approach or to talk or get agreements on diplomatic policy as well.

So, as I said, an understanding of technology these days is the luxury to be able to know how to use the devices and explore with ease scientific invention.

Infinity of IT (Information Technology) Diplomacy

To be able to understand each other or speak the same language, one will need the use of Information Technology to be able to minimize misunderstood words because of the application of IT Diplomacy.

We can overcome even more misunderstood words and understand one another. This is a civilization which also requires the achievement of reaching higher levels of reality for human production and exchange. This is to increase the happiness that no one knows yet when our society will reach the infinity of IT Diplomacy.

Development of technology in science is found to be one of the most vulnerable things to attack, meaning that one can come with invention and clearer applications.

Laboratories that will confirm your theories or inventions obsolete, and this can be done in a few seconds, is why a scientist must secure his theory or inventions with application standards of scientific methods: 1–Observation, 2–Analysis, 3–Hypothesis, 4–Prediction which out of this standard method your invention or scientific innovation can be even more vulnerable to attack, by other scientists, which is why science and technology will never be considered Utopia.

Believe me or not or consider me alone in my observation and experience of many years working in this most developed American scientific community. I can say that American society is the only nation in the world that reaches a percentage higher, the highest development in IT Diplomacy—as culture that attracts all the world, starts from great, universality, labs, Museums, i.e., in DC, Disneyland in California, and Disneyworld in Florida. Those are reality proof of how America is the most developed nation in the world based on our Information Technology Diplomacy.

Remember that to achieve 100 percent Information Technology Diplomacy, one must be able to maximize the potential of technology and science so that life expectancy or longevity will be at least a one hundred percent sphere of life (different areas of activities—

examples: family, friends, education, work, dating, career, and healthy travel.)

Endeavor must be one hundred percent, not just in living well but also a scientific community must reach a capacity of predicting incidents—phenomenon, as well as unintended consequences must be eliminated or detected in the instant of probability. There will be no limitation on human intelligence. Now, so far, as humans, we do not know enough.

I think that when any government is one hundred percent integral it means that society is a type of angel, therefore that society will not need a leader which means society is organized 100 percent. This can be considered the most compelling argument about integrity and diplomacy that I am endorsing for the new century. I think we have only one percent so far of IT Diplomacy in the world, in America and I must say that so far our government is also the unique most highest integrity above any nation, because IT Diplomacy increases a nation's control efficiency, and protection e.g., fire walls or Cybernet protection, law enforcement and most importantly our military and FBI, in daily basic practice on protecting business and government. We must continue developing it on a daily basis. Once again, I must include a formula to calculate integrity, as our security is never 100 percent, so let me borrow this phrase "trust but verify"/ Ronald Reagan.

May the food or way we eat be a constituent of diplomacy?

Of course, one who knows how to eat well and eat healthily can draw people to his niche or to his universe. Eating well is evidence of living healthy and that implies longevity, if we are acting intelligently. In this new century, employee information technology in the food industry such as biologic digital labels will help food buyers make smart choices and retain health. This is what I call "eat/educated by digital food menus".

This type of practice stimulates people to look up better dishes for healthy eating.

It is important to stabilize a healthy way of eating by following the next table, because we all know that healthy eating is part of what is needed to build our IT Diplomacy.

One must know how to eat, to at least consider simple math without extrapolation: there are a total of 20 amino acids; 11 that your body produces and nine amino acids that are the essential amino acids your body does not manufacture. The nine essential amino acids are: Histidine, Isoleucine, Leucine, Lysine, Methionine, Phenylalanine, Threonine, Tryptophan, and Valine.

These are the amino acids:

Fig-1

Essential	Conditionally Non-Essential	Non-Essential
Histidine	Arginine e.g. [peanuts 20 amino acids in variable proportions, USDA 2014]	Alanine
Isoleucine	Asparagine	Aspartate
Leucine	Glutamine	Cysteine
Methionine	Glycine	Glutamate
Phenylalanine	Praline	
Threonine	Serine	
Tryptophan	Tyrosine	
Lysine		
Valine		

Fig-2

Amino Acid	Main Food Sources
Histidine	Soy protein, **Eggs,** parmesan Sesame, peanuts
Isoleucine	**Eggs,** soy protein & tofu, whitefish, pork, parmesan
Leucine	**Eggs,** soy protein, whitefish, parmesan, sesame
Lysine	**Eggs,** soy protein, whitefish, parmesan, smelts

Methionine	**Eggs**, whitefish, parmesan, smelts, soy protein
Cysteine	**Eggs**, soy protein, sesame, mustard seeds, peanuts
Phenylalanine	**Eggs,** soy protein, peanuts, sesame, whitefish
Tyrosine	Soy protein, **Eggs**, whitefish, Smelts, sesame
Threonine	**Eggs**, Soy protein, whitefish, Smelts, sesame
Tryptophan	Soy protein, sesame, **Eggs,** winged beans, chia seeds
Valine	**Eggs**, Soy protein, parmesan, sesame, beef

CHAPTER 4

Is Ebola an Outbreak/ or Epidemic?

The Photo shows fellow Scientists: **Dr. James** and **Mario Nabliba** in charge of the Ebola Research Project.

LET'S LOOK AT WHAT is the need for safety and research to find a universal vaccine for Ebola?

Evidence of Ebola in West Africa in direct quoting:

"Ebola case # 3, which was treated and later died at Ngaliema Hospital in Kinshasa, Zaire. Ebola hemorrhagic fever (Ebola HF) is severe, often fatal in humans and nonhuman primates (monkeys, gorillas and chimpanzees) that have appeared sporadically since its initial recognition in 1976.

Ebola virus disease (EV), Ebola hemorrhagic fever (EHF), or simply Ebola is a disease of humans and other primates caused by Ebola virus.

Symptoms begin two days to three weeks after contracting the virus, with fever, sore throat, muscle aches and headaches. Normally, vomiting, diarrhea and rashes follow along with decreased liver and kidney functions. By this time, the affected people may start bleeding both within the body and externally.

The virus can be acquired in contact with blood or body fluids from an infected animal. Spreading in the air has not been documented in the wild. Fruit bats are believed to be a carrier and can spread the virus without being affected. Since human infection occurs, the disease can spread among people, too. Male survivors may be able to transmit the disease for nearly two months. To make the diagnosis, usually from other diseases with similar symptoms such as malaria, cholera and other viral hemorrhagic fever are ruled out first. To confirm the diagnosis, blood samples are tested for viral antibodies, viral RNA, or the virus itself.

Prevention includes decreasing the spread of disease from infected animals to humans. This can be done by checking these animals for infection and killing, and proper disposal of the bodies, the disease is found. Properly cooking meat and wearing protective clothing when handling meat can also be helpful, as are wearing clothes and washing hands protection when around a person with the disease. Samples of body fluids and tissues of people with the disease should be handled with special care.

No specific treatment for the disease is yet available. Efforts to help those who are infected are favorable and include giving either oral rehydration therapy (slightly sweet and salt water to drink) or intravenous fluids. The disease has a high risk of death, killing between 50% and 90% of people infected with the virus. EVD was first identified in Sudan (now South Sudan) and the Democratic Republic of Congo. The disease often occurs in outbreaks in the tropics of sub-Saharan Africa. From 1976 (when it was first identified) to 2013, the World Health Organization reported a total of 1,716 cases.

The largest outbreak to date is the course in 2014 with a West African outbreak of Ebola, which is affecting Guinea, Sierra Leone, Liberia and Nigeria. As of August 26, 2014, 3,069 suspected cases resulting in the death of 1,552 were reported. Efforts are underway to develop a vaccine; However, there is still no vaccine. This is one of the reasons why I have volunteered for research about an Ebola Vaccine.

[http://en.wikipedia.org/wiki/Ebola_virus_disease#mediaviewer/ File:Diseased_Ebola_2014.png]

CHAPTER 5

Research and Development (R&D) on scientific "Analysis on reasonable economy"

WITHIN THE SUPPORT PROVIDED by Russia to Guinea Bissau there were some betrayals admitted by same peoples of Guinea Bissau. As a matter of fact, the peoples of Guinea Bissau have said that since the long years of independence: stolen fish and corruption, just to name of few, witnesses to this in Guinea Bissau, claimed that in exchange for their defense; many long nights in 1980, one could hear sounds of cars passing, tractors and tanks. If those were tracks with construction materials the country would have fair capitalism, because those would have leveraged good public construction such as building good schools, roads, infrastructure, and department stores. Guinea Bissau could have been earning good GDP rather than the poverty today. There is another truth to admit that this nation has been the victim of their own arrogance e.g. of saying "bo leba tudo, pabia tudo na bing de conakry" this is translated to: arrogant people saying to Portuguese "you take all that you have because we have everything in Conakry." This was untrue and created destruction of the transition to the economy of the country.

As an American scientist of origin da Guinea, I predict that the formula below is the true application to apply:

IDPFEC (Integrity Data Protection Forensic Economic Country)=Land(Pasitive Economic)+"(Tax)"+ Exportation (Agricola) ^2- Government Expense+Technology/education

Therefore: IDPFEC=L+"T"+E^2-G+T/E

Higher taxation at any time can be problematic for economic circumstances, i.e., when taxpayer income is challenged, the taxations should be banned. Tax (T) application should be little or giving the tax stimulus, as our former President George W. Bush did in 2000 which helped American workers boost their production by investing this money back into our economy.

Forensic Economy for safeguarding of your production:

Integrity Economy and IT Formula to Safeguarding:

$$IDPFEC=L+"T"+E^2-G+T/E$$

As computer scientists: the kind of diplomacy that I am referring to is more focused on the IT area or so-called IT Diplomacy for the new trend in computer technology. Most importantly, we need to embrace the development of technology in use of digital media, social media, knowing the big picture of this is that Information Technology Diplomatic Venus, the practice of all good human behavior, is why it must embrace integrity and peace.

It is a unique way of performing business where scientists or entrepreneurs take advantage of most high-tech diplomacy to network with others rather than commuting to the workplace.

Organizing or providing a meeting online has created a backbone use of leadership as a must for IT Diplomacy.

There are many protocols and procedures of IT Diplomacy that must be used to make sure one can deliver the IT application without fear and for a clear subject to be increased to those that want the use of technology.

To not harm others, means the bad guys should be absent from access since day one that has been screened to that profile—of injurious behavior. This will ease the work of our law enforcement or IT Diplomacy can be used in the national pool for voters; that is to educate and create integrity in individuals to be honest and have no fear of how to vote.

One is taking the value or real character of IT Diplomacy in the use of technology. And this is what gives a full meaning to the American scientist who is prone to invent it (IT Diplomacy) so there is the complex of this subject that one must carefully analyze to understand the pragmatism of this book—*The American Scientist and Diplomacy*; it must be deliberated and deployed as part of the integral new science IDPF (New Trend), the leadership is the vector direction that one must use to help achieve it as a product. Human endeavors i.e. science must be self-evident based on four scientific methods:

Observation, Hypothesis, Analysis and Prediction.

This is to test and help to find a most economic success, the intention of most scientific inventors, it is important to recognize that the pragmatism of the inventor or scientist is to provide clear vision of IDPF which is a part of IT Diplomacy for use in leadership values that can embrace many areas such as Physics Biophysics, Computer Science which is the background of this author, in addition to interdisciplinary application.

How many of us will know, for example how to monitor COVID- 19, Ebola or any epidemic diseases if one does not provide the leadership skill required by scientists to control these diseases?

To project our leadership diplomacy in the scientific community, a scientist or diplomat needs to rush into the scientific invention of a worldwide vaccine that can be used for most wide spheres. This antibody will embrace and minimize worldwide diseases. Let's use our "IDPF DIPLOMACY" [INTEGRITY DATA PROTECTION FORENSIC DIPLOMACY], to make this clinical study achieve this purpose; this is the work to evidence what is narrative today in this pragmatism of an American scientist, which is to encourage again that with IT Diplomacy, we can make it a reality and like today with the existence of drones is a big revolution in the aviation industry. I am sure that one day drones will deliver the UNIVERSAL VACCINE to far points of the globe to treat patients as well as use of this IT, can take place now on many treatments helping virtual clinics or digital clinics and hospital work., that can be done with integrity or leadership diplomacy.

CHAPTER 6

Manufacturing, Business Requirement

THE MOST IMPORTANT ASPECT for making the project a success is to provide clear requirements. In order to accomplish this goal, the team must go out and complete a survey that encompasses system end-users and managers. There are several ways to conduct this survey, but our project will be concerned with Rapid Application Development (RAD). RAD is a group-based development plan with four phases: requirement planning, user design, construction, and cutover.

The Project Team will consist of Project Leader, System Engineer, two Technical Subject Matter Experts (SMEs), four system technicians, and the Company Information Technology Manager.

REQUIREMENT PLANNING

Before the beginning of the project, there will be a kickoff meeting to discuss the objectives and the schedule of the project. One of the objects will be the fact-finding phase in which the two Subject Matter Expert (SMEs) will conduct a survey on the new database install. While this task is being done by the SMEs, the System Engineer will consult with Oracle, and write a procedure in backup of data, migration of data, and install of latest innovation e.g. Oracle

Database, in the meantime, the four technicians will procure and design the necessary hardware used for the Oracle system. All the processes within this requirement planning are not "set-in-stone", and any changes will need documentation and approval by the company managers and Project Leader. Upon the completing of the "kick-off" meeting, the technical team will meet to discuss the design and development of the system install.

USER DESIGN

The System Analyst and Project Engineer will use CASE tools to achieve and complete their design phase. They will consult with end-users to gather more facts about how the system will be used and manipulated. With the new database installed, the system will be combined into one and provide data access using a web portal to internal customers as well as external customers.

CONSTRUCTION

The System Technician will procure and install the new hardware for the new database installation. Any necessary component will be acquired by submitting a Request to Purchase. This is the first stage of the procurement of the hardware. Only when the approval of this document is accepted that the purchase order can be drafted and the hardware ordered. When the parts arrive, the technician will assemble the hardware and install the new database install. After the installation, the Analysts and Engineer will configure and program the new software.

CUTOVER

After the initial migration of the old database to the new database installation, proper care must be taken to protect the integrity of the legacy database and its tables. The backup solution will be implemented at this time to keep the data in case any failures occur. The current legacy system will be kept online while the database is being tested to keep end-users productive. There will be a duration of 10 working days to fix bugs on the new system. At the end of the 10th day, the legacy will be offline, and the new system will take over its place. It is imperative that the legacy system remains intact just in case the new database install decides to fail.

RECAP

It is important to keep the project on schedule. In order to accomplish this task, everyone involved must stick to the "game plan". By providing a clear requirement and accurate schedule, the project will be complete in a timely manner.

Organization Manufacturing

Process Flow Charts, Procedures and Policy Statements

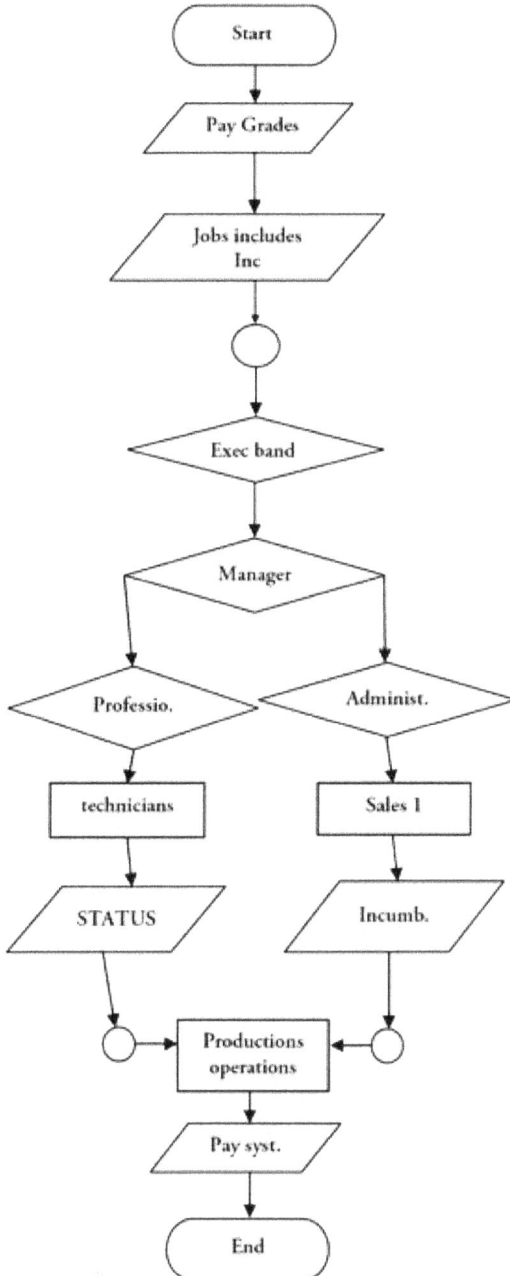

The meaning of some of the words in the flow chart are as follows:

Exec.   (Executive)

Professio.   (Professional)

Administ.   (Administration)

Incumb.   (# of Incumbents)

Syst.   (System)

Human Resources Procedures and Policy Statement

Organization Manufacturing's HR system is a constituent part of the financial systems package that is designed to keep track of the following employee information: payroll information, W4 forms, when the person was hired, their department information and seniority, and all personal contact information.

The organization is oriented with systems development where the following people could be involved: Information Systems Manager, Project Manager, Systems Analyst, Programmer, Business Manager, and end users. HR is in charge of making sure the personnel needed are on the job and are trained for it. Each personnel follow the procedures or series of steps of high tech (medical stents, heart valves, etc.) in the manufacturing industry. The HR department inspects the key elements of performance, interpersonal skills, and standards of ethical practice for business development issues. Training and Development records are kept in an Excel worksheet. This part may need to be complimented by an Oracle system training process which focuses special attention to employee training, where one has a "learning path that identifies the required courses for a desired training goal or certification level. There are recommended learning paths for job roles and one's path based on the Oracle product one wishes to learn." Also, we recommended some of the following requirements: that without the training no system can be successful without proper training, because of the successful information system requires training for users, Managers, and IT staff members.

Each recruiter maintains applicant information for open positions. Résumés are in a central storage area, and an Excel spreadsheet is used to track the status of applicants.

Workers' compensation is managed by a third-party provider, which keeps its own records. Employee files are kept by individual managers. There is no central employee file area. Managers are also responsible for tracking FMLA (Family Medical Leave Act) absences and any requests for accommodation under the regulation of the public facility, such as any portion of buildings, structures, site improvements, parking lots, or other areas of rescue assistance or evacuation elevators may be included as part of accessible means of egress on Riordan property. With the company regulations, the compensation manager keeps an Excel spreadsheet with the results of job analyses, salary surveys and individual compensation decisions.

Employee relations specialists track information about complaints and circumstances, regarded as just causes of protest, harassment complaints, etc. in locked files in their offices. The company has its own regulation system based on their philosophy which is divided in the following manner:

Annual pay adjustments—which comply with a process of annual performance evaluations in addition to pay adjustments, this activity occurs always yearly, the raises for employees also take effect the first day of their new fiscal year.

Management with the purpose and functionality of each Riordan team leader is based on the development activity of regulations which make sure the complete performance is demanded and must be followed. Every worker and administrator need a credible performance appraisal system. By using psychometric methods and computer technology, the process of developing such a system can be more efficient and demonstrably successful. Each employee who meets expectations or exceeds expectations, the managers are subject to gain compensations which is called" a poll of merit increase dollars. This procedure sorts many employees using the chart, where "x" corresponds to an average percentage of wage (salary) increase.

Regulations regarding reward practices are kept creating a team-oriented working environment. This also measures that communication occurs on a regular basis which supports and focuses everyone on the long-term viability of their organization.

The organization has factors in R&D (Research and Development) which are critical to the responsibility related to industry team leaders which clarifies the industry trends.

There is also a factor of ISO (International Organization for Standardization) 9000 standards which are very important to be considered in any organization because this is a standard that organization manufacturing agrees to where they use an auxiliary to receive *ISO 9000 certification*. This does not exactly mean that the organization's results obtained by performing have a high quality; "only that the company follows well-defined procedures for ensuring quality products."

The key employee relations are:

1. The organization has an "open-door" policy, where workers are free to share their opinions and are even encouraged to share their problems with higher-level supervisors if one was not happy with a management or supervisor action.

2. The employee received a handbook on the first day of hiring which contained the employee policies to follow such as attendance of organization.

3. Safety experts in the technical field or process are present on location to provide safety and health to those who need it in the work environment.

CHAPTER 7

IDP Forensic Equation and Pragmatic Definition

IN DEALING WITH ANY organization, one must consider where diplomacy fits into every aspect of it.

To work on creating this organization, I realize that life has a diversity of shapes in which I will describe with a thought; a leadership strategy to win must be based on the following: a vision without practice is a dead dream and practice without vision is a nightmare. It is fundamental that everything must be worked out in detail to achieve a goal because of the law of unintended consequences. Indeed, in all cases where life and human activity are involved, one basic need is to have a domain or control of any multifaceted activity. Even though the application of intellectual matter is not always dominant to the definition of pragmatism, in this case it is so Integrity Data Protection Forensic can be meaningful. The art of influence and directing those actions is done in such a way to obtain one's willing obedience, confidence, respect and cooperation. These confidences and multitasking are what we call leadership because it is built in a certain way to be acceptable to others.

For many, one might have an interpretation of their focus or be intelligent oriented to perform only one task. This is debatable, for me because Jasper Syndrome or Asperger's Disorder is a disease reflected in neurological disorders. It is considered a severe disease, one thought to be incurable, but IDP Forensic will argue this. Nothing is absolutely incurable so Jasper Syndrome can be a mild disease that has nothing to do with the exaggeration of a single intelligent oriented opinion.

This idea leads us to draw the line of this great organization, IDP Forensic, interdisciplinary and an activity to be able to exist and This is fundamental for the existence of any issue or science.

IDP Forensics' belief, research, practice and laboratory, can find the truth of anything. The dominant error that something is not curable is not also 100%; but it's ambiguities can be minimized. And so, it is a simple scientific procedure. IDP Forensic considers practice and research from debatable issues that we can demonstrate with the formula—

IDPF In $\int$ = $\int$ [total data - data vulnerable] x 100% / Total Data

The computation on this formula will never reach infinity; it tends to not be equal to 100%. It also means no security is absolute or equals 100% yet, because there are many hostilities that compromise our search to be true and absolute. Based on this idea that IDPForensic is founded to help and safeguard:

IDP Forensic (Computer Forensic Technology) is a safeguard and fundamental for any life endeavors.

Outsourcing the IT /Disaster Recovery

ProSpring Inc, my former organization, was founded by Jack Molisani in 1996. They advertise themselves by asking "Do you need an IT programmer, engineer, or IT project manager?

The technical staff division specializes in recruiting high-tech professionals, both contract and permanent. They can provide the right candidate with the right experience at the right price. Do you need a technical writer experienced in your industry? Outsource your IT writing project to their staff technical writers." (www. prospringstaffing.com)

What is Outsourcing?

Per definition, outsourcing is the buying of parts of a product to be assembled elsewhere, as in "purchase cheap foreign parts" rather than manufacture them at home.

However, this definition is more general than outsourcing information technology and is more about letting some experts in the field of disaster recovery be responsive to incidents in your organization which could be affecting your geographic area. If you are dealing with something outside of your country, it is called offshore. Indeed, outsourcing IT/Disaster recovery will be handled by an individual who has a domain strategy plan and is trained to reduce or minimize dangerous incidents. IT refers to resources such as computers, software, applications, hardware, and expert personnel as the main elements to control effectiveness of a project. Resource materials, as well as human labor are the first factors that determine the value of an organization by these things interacting with cost-effective plans for increasing the organization's efficiency which does determine one of the main factors that leads to a decision in proceeding with outsourcing.

With the use of information technology, the users could plan a way to increase the output or outcome of the plan to become more efficient according to applications. In a pinch, effective increases in an efficiency plan of the combats of computer disasters, for Outsourcing the IT/Disaster Recovery Function, for example, could be carried out with only a small-time investment and produce an immediate return. One could consider the following strategy: management of disasters is a straightforward organization of documents, plans, and drills of dangerous incidents in your workplace. The objective of outsourcing will be to find the best expert or professional to increase the ROI of the company. Managing outsourced work in IT is very popular today by offshoring your business to China or India rather than having the work done here at home because of the cost of living and commerce in the US. For example, South Africa or China, where the dollar's value is 8 times the Yen is a reason to offshore work. This economic difference stimulates the business owners to contract IT product support for disaster recovery outside the US. Nevertheless, the plan of action to pursue physical disaster recovery plans could make it impossible to offshore people.

Risks Associated with Outsourcing IT

Outsourcing IT risks will be convenient when one establishes communications by creating a Wide Area Network (WAN). In any business, an owner expects to make profits but there are always risks especially regarding outsourcing. This is often due to distance. For example, where different cultures are involved, there can be political factors at the destination or geographical location where the product is serviced. In fact, a WAN is needed for offshore outsourcing of departments of organizations. Offshore can be a negative factor in employment matters of a country due to political instability which can be a risk for the owner if any changes in a political situation result in loss of a contract. Regarding jobs being taken away to foreign countries, in this case, the organization who plans to offshore manufacturing jobs will find it is big business to China's benefit and the owner. Diplomacy is key here.

Outsourcing the IT /Disaster Recovery II

Contrary to popular belief, understanding the impact of offshore outsourcing does not require any formal economics training. Most economists and IT professionals assume the following ideal scenario when thinking about outsourcing. Offshore Outsourcing IT may be very risky at times when one confronts social-political, geographic and cultural situations. Social-political influences are a factor to consider in offshoring. Cultural difference in business ethics may create unexpected results in the understanding of tech support from one country to the next. Diplomacy is key again.

The Benefits Associated with an Outsourcing Effort

To be the most practical business nowadays, one must extend one's operations to many international locations where the need to negotiate a contract for services will take place to benefit the company owner because of the future projection of revenue. This practice is the new trend of businesses and applies to low-income countries, who charge low wages for their services. The most professional people are usually consultants who can do the work off site.

The costs originated in an outsourcing agreement, and examples of the dollar impacts expected can be more reasonable than creating a department in one's organization to handle that production. Good business management and according to agreements made is intended to cut costs by contracting the most expert consultants that can be found.

Let us pretend to create the following scenario:

> Before outsourcing the IT/Disaster Recovery, U.S. workers do tasks A, B, and C, and offshore workers are idle. After outsourcing the IT/Disaster Recovery, U.S. workers do tasks B, C, and D. Offshore workers do tasks A and some of B. In this scenario, U.S. workers were doing tasks A, B, and C and offshore workers were unproductive before the outsourcing occurred. After outsourcing, the U.S workers no longer do tasks A, but have a new task, D. Instead of there being an ideal scenario, the U.S. offshore workers are now doing tasks A and some of B so therefore the us workers remain fully employed but mixed in their tasks. Offshore workers are now fully employed with the tasks of A and B. This is what one hopes will happen.

There are two important problems with this scenario. First, the presumption is that US workers who were previously doing task A will easily be reabsorbed into the workforce by doing tasks such as B, C and D. This is what economists refer to as the "Adjustment" process. Unfortunately, the practical problems with adjustments are substantial. For example, let us assume that task A is computer programming for a certain software required in security, which is increasingly moving offshore, and task C, is testing hardware, which is in high demand in the us. This is realistic to accept. (Outsourcing America Ron Hira & Anil Hira, 2005)

Dollar Impacts That Might Be Expected

One does need the strategy of outsourcing and trade that give the ROI a boost. On a more fundamental level, because trade was never a problem of world development since the world began, divergent levels of development must trade using offshore means as the new way to do business. Indeed, the gain of outsourcing/offshore is tangible as long as there is a good prediction of potential danger in the negotiations that could occur between the two companies. What one needs to consider along with departmental care/management are those factors to minimize future risk.

Implications to the organizational business structure by using an outsourced IT department is a way the organization could extend their functions or enlarge their departments to operate in the worldwide capacity. The organizational size could develop to the size of a corporation that would need to create their own new IT departments. On the other hand, the security issue would be contingent upon how many resources the company must make every new area they create safe.

Regarding potential personnel issues which may arise, as usual in any organization, there are security issues that need to be minimized. Building teamwork is needed to have business intelligence. Human Resources will have to work very closely with hiring departments and security to make sure criminal elements are thwarted during the hiring process. Telecommunication (Network) is an important element of outsourcing/ offshore which needs to be established with security protection such as firewalls and encryption if needed. Business organizations must have auditing of a process and review of the use of their systems by personnel in the organization.

Outsourcing/Offshore IT

Sometimes in outsourcing/offshore plans there are cultural and language barriers which can influence the business even though the offshore team may have technical and skillful people to do the work. There will always be risk in any business to succeed as above but caution is used to thwart a disaster. To prevent an unfortunate situation, it would be better to review incidents of the past to study and prevent political barriers that could push aside an aspect of good outsourcing/ offshore and cause many business owners to not offshore their business. Those who outsource will make more profit by negotiating a business deal with better workers and less amounts to pay those workers which is considered cost effective or big savings.

Regarding outsourcing, here is an example of something that could happen; years ago, my organization relocated a technical writer to Paris for a company that also exists here in the US. This new contractor is American and found the cost of living very high in Paris. He then immediately asked to be transferred back to the US.

Indeed, I realize that it depends on geographic, political and economic situations of a country and this can play a big role for a company in outsourcing/offshoring to be successful with ROI and happy workers.

CHAPTER 8

Energy and the Environment

ENERGY AND ENVIRONMENT HAS been one of the focuses of my appreciation for our US Government therefore I feel my collaboration and research should be ongoing. I think this was one of the things I wrote about to my state senator.

Clean water and clean environments are necessary for our health. Greenhouse gases (GHGS) require our attention.

I think one thing we must do here in our American consumption of energy is to practice the use of energy efficiency. I want to mention the law and policy; nature, law and society are very important to follow and enforce if it is necessary to do so. Why? Because if planet earth is natural, it tends to be more environmentally friendly to humans.

This ought to be the work of social scientists, our lawmakers—our government regulations and scientists where both could benchmark applications of these policies. If we reward scholar programs that apply energy efficiency, it will help to decrease the use of detrimental energy that devastates our people e.g. hurricanes and other phenomena that we call nature phenomena. In my opinion, it is not natural because it is caused by an elevated temperature built on emissions of CO_2, it is scientifically proven—"a gas at standard temperature and pressure and exist in earth's atmosphere in this state".

To invent alternative energy, it must be done in a cost-effective way. One can apply axioms to sewage treatment, plants, and other pollution prevention and control systems. This is what our US government stresses in its plan/program "to reduce greenhouse gas emissions 80% by 2050."

This will indeed keep our lives safe and the environment healthy.

One must understand that there is accuracy of language needed when we are using these terms.

Climate and this also become a term now where everyone treats the climate as climate change, this can be the wrong term when we need to understand phenomenon behavior in each geographic area. I think we must be precise to disfiguring the climate with weather

What is the difference between weather and climate?
"Weather is the measure or assessment of the atmospheric conditions manifesting over a certain geographical location, at an exact moment in time. Climate is the measure or assessment of atmospheric conditions that manifest on a certain geographical location over an elongated period of time, often exhibiting repeatable patterns of change or stability on an annual or longer basis".
Reference: *www.nasa.gov/.../climate_weather.html*

The clear distinction of this is: Climate is long-term weather, opposed to weather as a short-term manifestation that we get in the United States, meaning short as a three-month period, opposed to climate, which is found to be extended time, approximately six months. This kind of extended period of time is to be considered the rainy season, and dry season, e.g. in other areas like African regions it is divided in six month rain and six month dry; meaning the rainy season started the 15th of May and ends in the first days of October and approximately 15 days posterior of October which used to be the beginning of the dry season, with less propagation of mosquitos. Therefore, I experienced the 15th of October to the 15th of May as the dry season in most African Countries or tropical climate. For one who will not be including some days or October 15th, September, November, December, January, February, March, and April to the first 15 days of May will still be the dry season. Exception is desert areas.

The scientific field, so called environmental engineering, must be applied to educational programs to reward more scholarship programs on the importance of those natural laws to prevent damage to the earth.

Scientific research or proofed use of alternative energy with cost effectiveness, can apply axioms to sewage treatment plants and other pollution prevention and control systems.

With stress in the plan, this matter greatly enables executing a program to reduce targetable greenhouse gas emissions and should be immediate for scientists as well as a mission for each of us as citizens and this action will indeed keep our lives safe from enemies of friendly environments.

I believe in the safety of our environment, so I give you a brilliant idea in quotes:

My humble idea:

> Individuals, without the cognitive: educational literacy, economics and lack of understanding of your culture, could decrease participation in physical and conservative environmental action as well as increasing rumors, because levels of knowledge determine your level of involvement in social justice and your interaction with the scientific community. By Mario Nabliba

And yes, our team of American leader entrepreneurs are the self-evidence of Information Technology Diplomacy, I must say that they are humble friends that have wisdom of what it will take for business and understanding of American people for their daily business proposals or most needed to employ leadership and development in the economy of organizations.

And I want you to know that Information Technology diplomatically has a via (one of the most important elements in this case is a wireless phone). A tool with a capability of connecting to the web and now one can live and breathe free because of Information Technology Diplomacy.

What is Diplomacy?

The term is a very fancy word because of the way it sounds and as a "project circuit answer" (I coined this phrase to mean a simplified word) which is to say an answer with one word–diploma.

It is but not exactly if there is no application of the real world to it. We must employ nature here naturally, when one deals with another and so one will be recognized as easy going and can change people's lives.

I want to give you an example of naturally employing life safety and trust to one, to be able to trust you:

On December 27, 2014, in California in one of my nearby cities, I pulled into a parking lot to go to Starbucks. In the first spot on the street, I saw someone behind looking for a spot. I used diplomacy to ask her if she wanted to have my spot behind me. She said yes so, I moved for her. This could interrupt my flowing time of production. My wife always said, "it is wise to not bother with non-important things, so please be ready to always use diplomacy to solve any problem because you will be a winner in many situations".

This reminds me to quote the important research I did in my graduate studies many years ago, reflection of a topic about employee law which I want to use as diplomacy, if you will, because scientists must understand a law to be able to endure their studies. I found a formulation made by Dr. David Abrahamson in the following equation: $C=T+S/R$, where C is Crime, T is Tendencies, S is Situations and R is Resistance (this means literally to not practice in bad situations). I ought to infer that R, for me, can be treated as Diplomacy. [(D) or (R=D)],

This formulation can be found as" CAUSATION, FACTORS, LAWS, PREDICTION." in a Book written by Walter C. Reckless, OHIO STATE UNIVERSITY

I think:

WHAT I REALLY WANT US TO TAKE FROM THIS IS TO UNDERSTAND THAT HE SHOULD ALWAYS USE DIPLOMACY TO RESIST A BAD CONSEQUENCE. I AM ASKING BY EMPHASIZING A WORD "PLEASE" TO USE IN OUR SOCIETY FOR DIPLOMACY TO RESOLVE ANY PROBLEM., SO WE SHOULD DO SO BY FOLLOWING THE AUTHORITY OF LAW.

High Topic: Diplomacy Is Always What Matters:

Diplomacy should be one of the most useful principles to apply to any conflict resolution.

In virtue of what I really want you to take from the real world is some of the practices of diplomacy; how one can utilize it to be able to solve many of life circumstances. I am a prolific writer, if you would Please use the standard legal definition of diplomacy. I am directly quoting before you agree with this; a definition from Black's Law Dictionary revised fourth edition defines diplomacy and I quote:

> Diplomacy: *Is the Science which treats the relations and interests of nations with nations. Negotiation or intercourse between nations through their representatives. The rules, costumes, and privileges of representatives at foreign courts.*

In conclusion I would like to say that diplomacy can be done; that means education, training and understanding of one another will even embrace diplomacy.

Information Technology Diplomacy was not efficient in the era of remarkable scientists: from Galiileo, to Dr. Luis Fona Tchuda, to us. Dr. Silva Natucan, a Physicist and I, in 1988 did not know the real use of diplomacy. There is somewhere a picture of me with my "diplomatic suitcase" (this is a misleading definition of briefcase if one translated this from the meaning in Portuguese "Pasta Diplomata" that anyone carrying a briefcase would be considered a diplomat). Seeing someone walking with just a briefcase does not mean anything. A real diplomat needs to have pure experience of Information Technology Diplomacy, academic knowledge of the real world, with international relations.

In 2012, I was at a meeting with my friend, International Diplomat, Dr. Diago Freitas do Amaral, where we talked briefly in Lisbon about diplomacy in Portugal and stabilized what I called US and Portugal Private Citizen Diplomacy. I remember there is a picture of us in Lisbon Portugal where I cultivated enough with him about these issues. (My heart goes out to my late friend Dr. Diago.)

For this new century that we are living in and the problems that we are facing on a regular basis it is required that each of us as citizens think.

One solution of these problems can be found in science and his diplomacy. This is what this book is about: American scientist and his diplomacy; we find science and our American Diplomacy to bring into the world development, these are our responsibilities as Americans and our responsibility as country leaders of the free world.

To be embraced as an American Scientist and Diplomat, one must fully understand our US most developed science technology that we have always known to be enterprise and innovative industries. Plus, the latest now known as AI.

I stood up with Suzanne Nabliba, my wife, in an airport going to a family reunion and watching a big airplane touch down on the ground and it made me think a lot about how amazing this nation is and how well it has done in technology.

And I feel like I want to be an ambassador for America, for me being good in technology I can then be a salesperson for this greater technology, for our foreign policy, to continue this great revolution of technology which will expand life, liberty, and the pursuit of happiness and most importantly to show to the world how good our culture is from our founder and protectors of America from the enemy, because "it can be done", again.

The Aircraft (DIAMOND DA40-G1000),
the ground training for safety.

Again, we need an Information Technology Diplomacy, to be able to succeed in anything.

If I was lacking in the skills of IT Diplomacy, I would not be a Lifetime Member of AOPA.

Pronouncements – Some of our greatest policy makers:

Policy Maker	Statement
Harry S Truman, 1945	"The United States should] take the lead in running the world in the way that the world ought to be run."
Dwight D Eisenhower, 1960	"My recent travels impressed upon me even more strongly the fact that men everywhere look to us."
John F. Kennedy, 1961	"Let every nation know, whether it wishes us well or ill, that we shall pay the price, beat any burden, meet any hardship, support any friend, oppose any foe to assure the survival and success of liberty. This much we pledge-and more."
Richard M Nixon, 1963	"I say that it is time for us proudly to declare that our ideas are for export. We need to apologize for taking this position."
Lyndon B. Johnson, 1965	"History and our own achievements have thrust upon us the principal responsibility for protecting freedom on earth…No other people in no other time has had so great an opportunity to work and risk for the freedom of all mankind."
Gerald R. Ford, 1976	"America has had a unique role in the world and ever since the end of World War II we have borne successfully a heavy responsibility for insuring a stable world order…We have taken the role of leadership."
Jimmy Carter, 1977	"Because we are free we can never be indifferent to the fate of freedom elsewhere"
Ronald Reagan, 1980	"We in this country, in this generation, are, by destiny rather than choice, the watchmen on the walls of world freedom

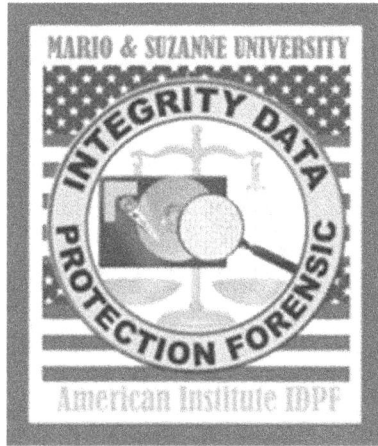

CHAPTER 9

Research on Vision Problems

MORE THAN A QUARTER of the world's population—some 2.2 billion people—suffer from vision impairment, out of which, one billion cases could have been prevented or have been left unaddressed, according to the first World Vision Report released by the World Health Organization (WHO) on October 8, 2019.

I, without any question, spend on average 13 hours a day consuming information or media on devices. Therefore, with the growth of Information Technology and the use of computers, cell phones and tablets, we need to be aware of the effects of light, or blue light, which can have a negative effect on vision without the propriety lenses for your vision protection.

In direct quote of America's most modern organizations, the department of Research, Development and New Trend of American Institute IDPF, we need to ask what blue light is and why is it harmful.

The prevalence of people that have distance visual impairment is **3.44%**, of whom 0.49% are blind and 2.95% have MSVI. A further 1.1 billion people are estimated to have functional presbyopia. Feb 8, 2018

In its natural form, blue light comes from the sun and is used to regulate the body's natural 24-hour sleep and wake cycle, known as the circadian rhythm. It helps boost alertness, elevate moods and generally makes people feel better. Blue light is also helpful in treating Seasonal Affective Disorder (SAD) and certain skin conditions. Artificial light emitted from the fluorescent light, LED light, flat screen TVs and tech devices are shorter High Energy Visible (HEV) wavelengths—and that is where the problems start.

A Harvard University study reveals that blue light suppresses melatonin more powerfully than regular light and affects the sleep cycle because our bodies don't know it is time to sleep, thereby throwing off the circadian rhythm. Researchers suggest turning off all devices two to three hours before bed; using a dim red night light which has "the least power to shift circadian rhythm and suppress melatonin;" and to get exposure to bright light during the day to help sleep better at night.

Blue light can also affect your eyes. It's no secret that looking at a screen for a length of time causes eyestrain and dry eyes, but according to optical chemistry research at the University of Toledo, "blue light from digital devices and the sun transforms vital molecules in the eye's retina into cell killers ... which may lead to age-related macular degeneration, a leading cause of blindness in the United States." In order to protect the eyes, the researchers state it is important to wear UV-protective sunglasses outside and to avoid looking at cell phones in the dark.

"There are many different types of vision problems that could be affecting your eyesight. But, for the purposes of this guide, we will be focusing on four of the most common causes of impaired vision. These, along with many other vision impairments, are treated with care and precision."

Astigmatism

Astigmatism is an uneven or irregular curvature of the cornea or lens, which results in blurred or distorted vision. Other symptoms of astigmatism include the need to squint, eye strain from squinting, headaches and eye fatigue.

In reality, most people have some degree of astigmatism, which is usually present at birth and is believed to be hereditary. In minor cases, treatment may not be required but is certainly beneficial. Moderate to severe astigmatism can be treated with corrective eyewear or LASIK surgery.

Hyperopia (Farsightedness)

Farsightedness, medically known as hyperopia, refers to vision that is good at a distance but not at close range. Farsightedness occurs when the eyeball is shorter than normal, as measured from front to back, or when the cornea has too little curvature. This reduces the distance between the cornea and retina, causing light to converge behind the retina, rather than on it.

If you are mildly farsighted, your eye care provider may not recommend corrective treatment at all. However, if you are moderately or severely hyperopic, you may have several treatment options available, including eyeglasses, contacts, LASIK and photorefractive keratectomy (PRK).

Myopia (Nearsightedness)

Nearsightedness, medically known as myopia, refers to vision that is good at close range but not at a distance. It generally occurs because the eyeball is too "long" as measured from front to back.

Nearsightedness is diagnosed during routine eye exams and possible treatments include eyeglasses, contacts, acrylic corneal implants, LASIK, radial keratotomy (RK) and photorefractive keratotomy (PRK). Your eye care provider will suggest the best treatment option for you.

Presbyopia (Aging Eyes)

Aging eyes, medically known as presbyopia, is a condition in which the lens of the eye gradually loses its flexibility, making it harder to focus clearly on close objects such as printed words. Distance vision, on the other hand, is usually not affected.

Unfortunately, presbyopia is an inevitable part of aging and cannot be prevented by diet, lifestyle or visual habits. However, it is treatable with several types of corrective lenses, including progressives, bifocals and trifocals, single vision reading glasses, multifocal contact lenses and monovision therapy.

Eye Diseases

"Eye diseases" is a blanket term that refers to a host of diseases relating to the function of the eye. Below we describe some of the more common types of eye diseases and how they are generally treated.

Conjunctivitis (Pink Eye)

Conjunctivitis, also known as pink eye, is an infection or inflammation of the conjunctiva—the thin, protective membrane that covers the surface of the eyeball and inner surface of the eyelids. Caused by bacteria, viruses, allergens and other irritants like smoke and dust, pink eye is highly contagious and is usually accompanied by redness in the white of the eye and increased tearing and/or discharge.

While many minor cases improve within two weeks, some can develop into serious corneal inflammation and threaten sight.

Diabetic Eye Disease

Diabetic eye disease is a general term for a group of eye problems that can result from having type 1 or type 2 diabetes, including diabetic retinopathy, cataracts and glaucoma.

Often there are no symptoms in the early stages of diabetic eye disease, so it is important that you don't wait for symptoms to appear before having a comprehensive eye exam. Early detection and treatment of diabetic eye disease will dramatically reduce your chances of sustaining permanent vision loss.

Glaucoma

Often called "the silent thief of sight," glaucoma is an increase in the intraocular pressure of the eyes, which causes damage to the optic nerve with no signs or symptoms in the early stages of the disease. If left untreated, glaucoma can lead to a decrease in peripheral vision and eventually blindness.

While there is no cure for glaucoma, there are medications and surgery available that can help halt further vision loss. Early detection and regular eye exams are vital to slowing the progress of the disease.

Macular Degeneration

Macular degeneration is a chronic, progressive disease that gradually destroys sharp central vision due to the deterioration of the macula, a tiny spot in the central portion of your retina comprised of millions of light-sensing cells. Because it is so commonly associated with aging, it is also known as age-related macular degeneration (AMD). There are two forms of AMD called "dry," most common and with no known treatment, and "wet," less common and treated with laser procedures. Genetic testing is now available to help identify those most likely to develop "wet" macular degeneration.

In most cases, reversing damage caused by AMD is not possible, but supplements, protection from sunlight, eating a balanced diet and quitting smoking can reduce the risk and progression of macular degeneration.

Eye Conditions

There are many different types of eye condition that could affect your eyesight or could have long-term consequences if not treated properly or promptly. We list some of the more common conditions below. If you think you or someone in your family has one of these conditions, encourage them to seek professional treatment.

Amblyopia (Lazy Eye)

Lazy eye, medically known as amblyopia, is a loss or lack of development of vision, usually in one eye. This degenerative process usually begins with an inherited condition and appears during infancy or early childhood. Lazy eye needs to be diagnosed between birth and early school age since it is during this period that the brain "chooses" its visual pathway and may ignore the weaker eye permanently.

Lazy eye is not always easy to recognize since a child with worse vision in one eye does not necessarily have lazy eye. Because of this, it is recommended that all children, including those with no symptoms, have a comprehensive eye examination by the age of three and sooner if there is a family history of any eye condition or disease.

Blepharitis

Blepharitis is a general term for inflammation of the eyelid and eyelashes. It is among the most common and stubborn eye conditions usually resulting from poor eyelid hygiene, a low-grade bacterial infection (usually staphylococcal), an allergic reaction and/ or abnormalities in oil gland function.

Like some other skin conditions, blepharitis can be controlled but not cured. The main goals in treating it are to reduce the number of bacteria along the lid margin and open plugged glands.

Cataracts

A cataract is a clouding of the eye's normally clear lens, which leads to a progressive blurring or dimming of vision. It is the world's leading cause of blindness and among the most common conditions related to aging—by age 65, you have a 50 percent chance of developing a cataract, and, by age 75, it jumps to 70 percent.

A cataract starts out small and initially has little or no effect on vision. As the cataract progresses, it becomes harder to read and perform other normal tasks. In the early stages, your doctor may recommend stronger eyeglasses and adjust your lighting to reduce glare. When cataracts disrupt your daily life, your doctor may recommend cataract-removal surgery, which is one of the most frequent and successful procedures done in the U.S.

Computer Vision Syndrome

Computer Vision Syndrome is characterized by neck pain, blurry vision, stiff shoulders, headache and watery eyes when working in front of a computer screen. The symptoms are typically due to posture, dry eyes, eye muscle coordination and poorly corrected vision.

Since computer monitors are typically 20 to 26 inches from your eyes, your regular glasses may not be the best option for computer work. This distance range is considered intermediate—closer than what you use to drive a car but farther away than what you use to read. Special lens designs for computer work provide you with a larger intermediate area for viewing the computer and your immediate work area like the top of your desk.

Dry Eye Syndrome

Dry eye syndrome refers to a breakdown in the quantity or quality of tears to moisten, cleanse and protect the eyes. This is significant because, with each blink, tears protect the surface of the eye, washing away dust and microorganisms. When this protective coating dries up, the eyes may feel "gritty" or burn and can be more sensitive to light. In extreme cases, vision can be blurred.

If you suspect that you have dry eye, see your eye doctor. Proper care will not only increase your comfort—it will protect your eyes. Your eye care provider can perform a series of tests to determine if you have dry eyes.

Strabismus

Cross-eyed, medically known as strabismus, refers to a condition in which eyes are misaligned. It commonly occurs when the muscles that control eye movement are not properly working together.

The result is one or both eyes turning inward, outward, upward or downward, or one or both eyes moving irregularly. Strabismus is usually diagnosed during childhood and affects about 4 percent of children, afflicting boys and girls equally. Though it cannot be prevented, its complications can be avoided with early intervention. Even if you notice symptoms intermittently—when your child is ill, stressed or fatigued—alert your eye care provider.

Vision Care & Products

- Lenses and Frames
- Contacts
- Vision Correction
- Eye Conditions
- Eye Diseases
- Vision Problems

Vision Correction

Vision correction is a general term used to describe a variety of optometric techniques for correcting less-than-perfect vision. For your convenience, we have included a brief description of some of the most common vision correction procedures.

Corneal Reshaping (Orthokeratology)

Orthokeratology is a procedure for correcting myopia (nearsightedness) and mild astigmatism by gently reshaping the cornea with special contact lenses, which the patient places in his or her eyes overnight.

When successful, patients will experience clear vision during the day without contact lenses or eyeglasses. However, the results are temporary, so the patient must continue to wear the lenses regularly at night to maintain optimum results.

LASIK

LASIK (Laser-Assisted in Situ Keratomileusis) is a surgical procedure that uses a laser beam to reshape the cornea. Patients who are nearsighted, farsighted or astigmatic may benefit from this type of procedure.

While millions of patients have seen successful results from LASIK, the procedure is not right for everyone. Your optometrist will need to thoroughly examine your eyes to determine which type of vision correction best fits your needs.

Low Vision Therapy

Low vision is a general term that refers to a partial loss of vision that cannot be adequately corrected with eyeglasses, contact lenses, medications or surgery. Common causes of low vision include macular degeneration, diabetic retinopathy, inherited retinal degenerative diseases, glaucoma and optic nerve atrophy.

Low vision therapy typically includes an evaluation of the patient's visual abilities, prescription of low vision devices and training in their use. The goal is to maximize the use of the patient's available vision for reading, writing, hobbies and work-related tasks such as working at a computer.

Vision Care & Products

Lenses and Frames
Contacts
Vision Correction
Eye Conditions

Lenses and Frames

Selecting your lenses and frames is a very personal choice. A wide variety of options are available that can be tailored to suit not only your medical needs but also your fashion preferences.

Highlight your features, play with color, and augment vision for different functions such as reading, driving and playing sports. With all the choices available, the experience can be daunting without the right guidance.

Eyeglasses

Design, material and treatments are the three components that make up a pair of prescription lenses. It is important to select the right combination of these elements for your visual needs and to always consult your eye care professional.

Eyeglass Lenses

Selecting the right eyeglass lens depends largely on its function, from single vision lenses to progressive polycarbonate lenses.

Regardless of your situation, your eye care provider can help determine what types of lenses will work best for you in terms of comfort, function and design.

Frame Styles

When choosing a frame, the shape and size of the frame should enhance the color of your eyes, complement your skin tone and play up the best features of your face shape.

Most people need more than one pair of glasses, such as one for everyday wear and another for outdoor activities. Having different style frames for different activities and moods makes wearing glasses more fun.

Sunglasses

With the wide variety of lens options available, you can customize your "sunnies" (sunglasses) to meet your visual, protection, performance and comfort needs. Sunglasses protect your eyes from harmful ultraviolet (UV) radiation, which is present even on cloudy days. Quality sun wear provides 100 percent UV protection and can significantly reduce the risk of vision problems caused by sunlight such as cataracts and retinal damage.

Glare, an issue that makes it difficult to see objects clearly by washing out colors and details, can be combated by polarized lenses. Looking at a scene with polarized lenses, you'll notice the colors are deeper, richer and bolder, and details are clearer and more distinct. Polarized lenses also help reduce squinting, which, in turn, reduces eye fatigue, tension and eyestrain.

Anti-Reflective Lenses

Wearers of prescription glasses and sunglasses commonly encounter annoying glare and reflections caused by light bouncing off their lenses. This glare makes it more difficult to see, especially at night.

Anti-reflective lenses reduce these reflections, allowing more light to pass through to your eyes.

All lens surfaces naturally reflect light, and this reflection can prevent between seven to 14 percent of the light needed for optimal vision. Wearing non-AR lenses is like trying to read a book in a dimly lit room. Since AR lenses allow more light to reach your eyes by reducing reflections, it's like turning up the lights in a room, making it easier to see.

Lenses and frames are a very necessary and personal choice.

I have been a patient of Kaiser Vision Care. I have experienced presbyopia defined as "the physiological decline in the eye's accommodative power, resulting in an inability to focus on near objects clearly." My eyes are less flexible because of aging and as a patient and scientist in my own research I experienced and made a major improvement to my eyesight by using "OCUPRIME"–the vision support formula with vitamins, minerals and herbs.

Reference: Amazon.com: IDEAL PERFORMANCE Ocuprime for Eyes Supplement Vision Pills (60 Capsules) : Health & Household

ACKNOWLEDGMENTS

I WOULD LIKE TO thank the many people who have helped me at various stages of research and work for this book. From graduate study—A Master of Computer Science in Information Systems (MSCIS), my pre-doctoral research in Biophysic statistics through various universities has been the journey this book describes and has been hard but sometime fun. I have met many academic professionals from the following: universities, UCLA, USC [Los Angeles], UC Berkley [California], Harvard, University of Arizona, UOP, UT [Austin Texas], Bradley University [Peoria, IL] University of Birmingham [UK] and University of Lisbon [Portugal], Cal Tech Institute of Technology [California]. ALL THESE SCHOOL I AM thankful for their support and resources.

I must stress how Suzanne and I enjoyed the work trip that led us to East Coast Harvard in Boston and staying at the Double Tree Hotel was such a joy to be able to explore some of the food that is a reality now in this book.

I want to thank and have been grateful for my superiors at the entire academy of the Veterans Affairs Greater Los Angeles, for their major resources and labs during my work as a staff scientist for 10 years. I acknowledge that I have met and worked with many top scientists.

I extend my grateful acknowledgement to my late parents: My dad had experiences during World II and kept memories and shared this history of the Portuguese in the 1940s. I want you know that my dad barely read or wrote but was able to produce several empirical details of what was a great time in "Guinea Ultramarino" or Guinea Portuguesa now known as Guinea Bissau. There was good wine, great Portuguese living life, that he never forgot. He worked for the Portuguese Ultramarino and met Portuguese peacekeepers and some NATO friends who were providing healthcare and vaccination in the region. Thank you, dad for teaching us Portuguese at home and sending us to Portuguese schools, and mom who later became a USA green card holder and loved America and American people. Americans loved her back. She call my wife Suzanne her golden beautiful Americana, as she said in her own word.

Another special thanks go to our federal system of student loans that I was able to pay off. I also thank the special people of my family and in-laws: kudos to my lovely wife, Suzanne Elizabeth Crowley Nabliba, former Vice President of ProSpring Inc, Board and Executive Advisor of Mario and Suzanne University American Institute IDPF, who worked tirelessly for this book. Thanks for her understanding in letting me work such long hours in the career of a scientist that I love. Thanks to my sister-in-law, Maureen Clark, who retired from the FAA and worked on reviews of this book. Thanks to my brother-in-law, Tim Crowley, a brave veteran of the Vietnam War. He has a love of computer graphics and helped me with my work. He needs to be recognized as a veteran. He left work one day for lunch and when he came back his workspace had been bombed by the enemy. Thank God my brother Tim was safe and alive. He is 80 and still does great graphics. Thank you for your service.

I must thank Mr. Michael Reynolds, long term friend, very funny man. Michael is a famous Hollywood movie director, and someone you can rely on. He and I worked on putting some parts of this book into a movie script with integrity. We hope it will be a movie one day. Thanks to him very much, indeed for his hard work and partnership in Hollywood California.

REFERENCES

Chapter 1: American Institute IDPF www.idpforensic.om

Chapter 2: Tracking and Managing Our Health lead to the heart. (It's [California] Genomics at University of California, Davis). [Ruth Debusk, PhD, RD and Yael Joffe, RD(SA),2006, pp.26-28.]

Chapter 3: Washington DC Research: *http://www. usatoday.com/story/money/business/2014/04/14/ coca-cola- coke-freestyle-soft-drinks-beverages/7478341/*

"trust but verify"/ Ronald Reagan

Chapter 4: Topic: Is Ebola an Outbreak /Ebola or Epidemic? Need for safety, Food safety find a Universal vaccine for it.

[http://en.wikipedia.org/wiki/Ebola_virus_disease#mediaviewer/ File:Diseased_Ebola_2014.png]

http://www.cdc.gov/vhf/ebola/resources/virus-ecology.html

http://en.wikipedia.org/wiki/Ebola_virus_disease#mediaviewer/ File:Ebola_virus_virion.jpg

http://en.wikipedia.org/wiki/Ebola_virus_disease#mediaviewer/ File:Diseased_Ebola_2014.png

Chapter 5: Research and Development (R&D) on scientific "Analysis on reasonable economy"

Chapter 6: Manufacturing, Business Requirement

Chapter 7: IDP Forensic Equation and Pragmatic Definition

This formulation can be found as "CAUSATION, FACTORS, LAWS, PREDICTION." in a book written by Walter C. Reckless, OHIO STATE UNIVERSITY

Outsourcing America Ron Hira & Anil Hira, 2005

Chapter 8: Energy and the Environment

This formulation can be found as "CAUSATION, FACTORS, LAWS, PREDICTION." in a book written by Walter C. Reckless, OHIO STATE UNIVERSITY

Chapter 9: Research on Vision Problems

https://www.downtoearth.org.in/health/more-than-a-quarter-of-the-world-s-population-has-vision-impairment-who-67147https://www.health.harvard.edu/staying-healthy/blue-light-has-a-dark-side

https://orangetwist.com/kind-of-blue/

https://www.ncbi.nlm.nih.gov › articles › PMC5820628

https://wattersvisioneyecare.com/

ABOUT THE AUTHOR

MARIO NABLIBA, SCIENTIST

Dr. "Lincoln" MARIO NABLIBA, Scientist

PROFESSIONAL EXPERIENCE

Executive–Chief Information Technology/Scientist, Chief Executive Officer Founder, CEO and Chief Financial Officer (CFO) of M&S University American Institute IDPF since April 4, 2017, to present. I led IT operations and technology management, earned a Private Investigator (PI) from our honored USA Company – NSF.

Trained and Certified for Legislation Chief Judge: Served Federal and State election at the State of Maryland, Mario Nabliba was able to save an Electronic Voting System that failed to continue functioning in the middle of voting day. Mario used his electronic expertise and leadership as Chief Judge by troubleshooting the problem and bringing back the functionality properly. This action amazed American voters regarding Dr. Mario Nabliba's leadership.

EXPERIENCE:

IT Medical Database Manager:

Maintenance and Administrative Hardware and Software for patient medical records, employer and employee email systems and in house server (intranet).

Greater Los Angeles Veterans Healthcare System Staff Scientist in Department of Research and Development:

Accomplished a new database sheet for management of chemical inventory.

> "The consensus of senior staff of the Biosafety Committee was that this was an impossible task that could not be done by anyone and would never get accomplished". I did it.

Department Oncology: Assist with Administrative work for the Physicist Project Manager

System Analyst:

Troubleshoot and maintenance of PC Systems (hardware, software, and data) processed data for people's business needs Networking, maintenance of databases, Database Management.

Troubleshoot and maintenance of PC Systems (hardware, software, and data) processed data for people's business needs.

Networking, maintenance of databases, Database Management.

Utilizes project management methods to ensure projects are planned appropriately and completed within established guidelines. Coordinates and conducts staff training in-house system and communications systems. Supervised Service Desk staff, ensuring a high level of customer service and end user support; provides back-up support for the Service Desk and resolves escalated incidents; provides emergency support for system outages disasters.

STRENGTHS: Hard worker, ethical, in-depth knowledge of government and VHA Computer Network system.

NIST [National Institute of Standards and Technology]

Official invitation of International Conference: with NIST

Mario is recognized as the CEO of IDPF and this professional training and educational conference: that addressed the following agenda:

ADVANCED PROFESSIONAL RESEARCH, CONFERENCE AND PROGRAM OVERVIEW:

XCCELERATE Advancing Cyber Innovation for Government. Wash. DC NIST and the Forensic Science Center of Excellence – CSAFE (Center for Statistics and Applications in Forensic Evidence)

Forensic Science, Research Program Manager, Federal Program Officer, NIST

Distinguished Professor, Director, CSAFE STATISTICS

Overview of Work in the Statistics Forensic Science Focus Area at NIST and Applicable CSAFE Research

A Measurement Science Perspective on Assessing the Weight of Evidence {presentation will be made available upon completion of editorial review}

Measurements and Scoring Procedures for Footwear Impression Comparisons

Statistical Models for the Generation and Interpretation of Shoeprint Evidence

Modeling the Distribution of RACs (Accidentals) in Footwear Evidence

CSAFE-CMU DIGITAL AND IDENTIFICATION FORENSICS Digital and Identification Forensics

Quantifying the Weight of Friction Ridge Evidence: Score-based Likelihood Ratio for Fingerprints

New Data Collections in the NSRL: The NSRL Goes Mobile

StegoDB: An Image Dataset for Benchmarking Steganography Detection Algorithms

CSAFE-ISU

Statistical Methods for Change Detection Over Time in Digital Forensics Data

CSAFE-UCI ILLICIT DRUGS AND TOXINS

Emerging Designer Drugs and Synthetic Marijuana

Gas Chromatography Mass Spectrometry (GC-MS) Libraries for the Identification of Controlled

A Better Understanding of Cannabis Chemistry to aid in Vapor Phase Detection

NMR in Forensic Drug Analysis Statistical and Algorithmic Approaches to Matching Bullets

CSAFE-ISU

Developing Methods for Comparison of Cartridge Breechface Images

CSAFE-CMU TRACE EVIDENCE

Trace Evidence Measurements and Standards

Assessment of a Portable Spectrophotometer for Measuring Color of Automotive Paint Trace Evidence

Evaluating Sources of Variability in Forensic Fiber Trace Evidence Examination

NIST FORENSIC GENETICS

Applied Genetics

Interpretation of Complex DNA Mixtures

Population Sample Sequencing at NIST

Development of the Next Generation of the PCR-Based DNA Profiling Standard: SRM 2391d

NIST CSAFE HIGHLIGHTS

Latent Fingerprint Proficiency Testing

CSAFE-UVA

Combining Fluid Dynamics, Statistics and Pattern Recognition in Bloodstain

Pattern Analysis, to Quantify Spatial Uncertainty and Remove Human Bias

CSAFE-ISU

Human Factors in Identification Decisions: Cross-cutting Interdisciplinary Research

Materials Deposition Inkjet Printing for Spatially Resolved Chemical Standards of Relevance to Forensics

TEACHING EXPERIENCE

1990 – 1992 [Secondary School–Olave Moniz, Spargos Sal, Rep. Cape Verde]

Lecturer – Teach Math at High School level Developed syllabus and overall course structure, and administered 7, 8, 9 grades.

Liceu Regional–2 Agostinho Neto, Bissau Rep. Guinea Bissau

Developed syllabus and overall course structure, and administered 7, 8, 9 grades.

Liceu Regional – de Farim, Rep. of Guinea Bissau – Chairperson-Coordinated Physics

Developed syllabus and overall course structure, including lab practicum, and administered all grades [7, 8, 9] Collaborated on curriculum and exam development, met with students and parents upon request, planned the written work, including final exam papers.

1988-1998-Liceu Regional–1, Bissau, Rep. of Guinea Bissau–Instructor Apprenticeship – Physic Sciences developed syllabus and overall course structure, and administered 7, 8 grades.

RELATED EXPERIENCE:

Expert in Biophysics Study

1995 to Present Computer Scientist, PI (Principal Investigator) recognized by US/ NSF (National Science Foundation).

Researcher–with experience on Research and Development

NutraMetrix (Advanced Nutraceuticals)–Consultant/Certified / Products Market America

RESEARCHER – IMPORTANT RESEARCH

July 27–29, 2007 Harvard University, Computer Information Systems – On encryption and decryption

Feb. 16–23, 2007 Faculty of Science University of Lisbon and Italy

Feb. 07–10, 2007 University of Birmingham, England – Computer Forensics Feb. 21, 2006 Berkeley University, CA–Regarding Informatics Collaborative

Research Network (CRN), across campus, spanning the sciences, engineering, physics and medicine departments.

The primary focus of the research activities of the Informatics CRN are the following themes: modeling and analysis of large complex systems, knowledge bases, data analysis and visualization, and collaborative research technologies and infrastructures, during my Master's Program.

Scholarship financed by the Foundation of Calouste Gulbenkian Lisbon

1993 research at Institute of Physic Galileu Galie (departimento di fisica Galileo galley)

Padua Italy. Physic Science Program at Faculty of Science University of Lisbon

11/2012-PUBLICATIONS AND PAPERS

Integrity Data Protection Forensic [Computer Forensic Technology] New Trend

This book is Certified by Library of Congress ISBN: 9781479725502. 2011

[Managing Individual and Patient Performance]

Paper distribution to doctors to help patients managing 2010

Business speaking Aug. 12, 2007, Renaissance Speakers Hollywood, CA Celebrity Center International. Importance of use of information systems for business and using advanced presentation skills, speech about business on 4 quadrants of business: small business, self-employed, employees, investor and big business.

EDUCATION

University of Phoenix–La Miradas, CA

MCSIS [Masters of Computer Science Information Systems]

"Faculdade de Cience Universidade de Lisboa]/Portugal"

Physics /Post BA in Physics/Chemical [Minor: Biophysics] "Instituto Normal Superior Tchico-Te/ Guinea Bissau Areas of Concentration: Physics Sciences"

LANGUAGES

English–speak fluently and read/write with high proficiency

Portuguese native language

Spanish–read and write with basic competence

French–speak,

Italian–speak and read

MEMBERSHIPS AND HONORABLE DOCTORATE

Awarded a Certificate of Appreciation by Veterans of Foreign Wars of the U.S 2006 [US Foreign War Veteran -Award]

A Member of American CTC (Convergence Technology Council)

A Member of American CTC (Convergence Technology Council)

Elected Vice-President of the Board of Directors, Monterey Hills Federation

Friendship of South Pasadena Public Library

2018–U.S. Senator Candidate for California 2018 National Candidate Choose by partisan. 2019–HONORABLE DOCTORATE: Honorable Doctorate by my professional career at M&S University American Institute IDPF, Masters in Computer Science in Information Systems, STEM work in addition to highly recognized works in the field of Science and Technology [i.e. interdisciplinarity Computer Scientist's research breakthrough infection to organic beer, groundbreaking work as Staff Scientist at honorable Veteran Hospital in Los Angeles, CA.

Certified achievement completed, presented by AOPA AIR SAFETY INSTITUTE of hard work, study recognized. Scientific achievement of Navigating Today Airspace.

2022–Honorable position obtained through the tedious training and examination per legislative law of the State of Maryland and federal government bestowed upon me to serve as, "Chief Judge" as the state designation for this leadership position, where I managed a team and safe guarding, valuable technology of data during federal and state election.

2008 to Present: CEO and founder of Mario and Suzanne University/American Institute IDPF ONLINE SUPPORT

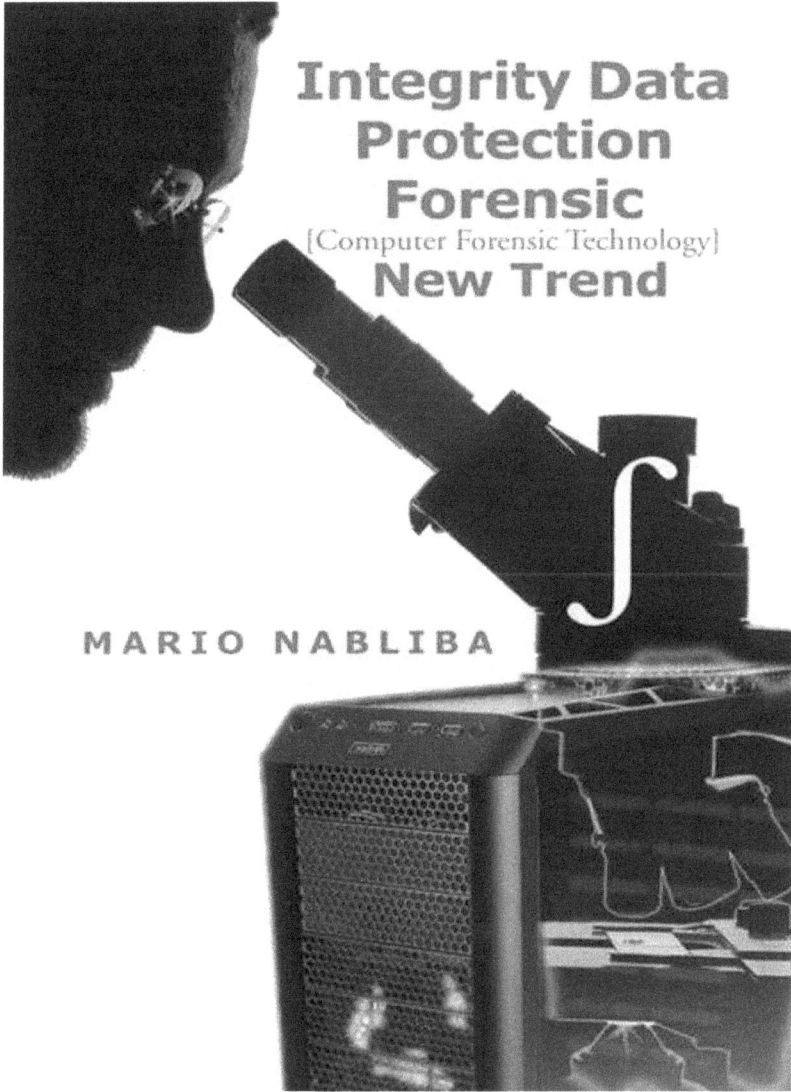

**Integrity Data
Protection
Forensic**
[Computer Forensic Technology]
New Trend

MARIO NABLIBA

🖋️ INVENTHELP

Dear Mr. Nabliba:

Enclosed is the Basic Information Package report, which you have requested for your invention, the "INTEGRITY DATA PROTECTION (IDP)". In this report, we have provided the service of "packaging" you invention; that is, we have assembled basic information relevant to the "INTEGRITY DATA PROTECTION (IDP)" in an organized report format that can serve as a handy reference tool.

Primarily, the Basic Information Package report is a resume of the "INTEGRITY DATA PROTECTION (IDP)" summarizing its positive and most appealing features, just as a resume assembles the assets of an individual seeking a job. As you will recall from our Services and Fees Flow Chart, InventHelp® also performs a submission service under a separate contract. If you decide to proceed with our submission program, the Basic Information Package report will serve as the basis for the preparation of descriptive materials which will be presented to industry in the hope of obtaining a good faith review of the "INTEGRITY DATA PROTECTION(IDP)". If you decide to promote your invention on your own, the Basic information Package report can be a useful reference, and it can also be used by you to stimulate interest among potential investors.

Our submission agreement will permit us to present the "INTEGRITY DATA PROTECTION (IDP)" to industry and review any interest that may be expressed. We look forward to working with you.

Research Department
InventHelp

Description of Duties/Responsibilities:

Mr. Mario Nabliba first came into Research Service on a volunteer basis in mid December 2009 as was introduced to me for a special assignment that I had developed. That special assignment was essential for the development of a new database for the Research Service management of the chemical inventory.

The work I assigned to Mario involved researching very specific information on the hazardous properties of each of several thousand chemicals in our inventory and enter that information into a developing database. The consensus of the senior staff on the biosafety committee was that this was an impossible task that could not be done by anyone and would never get accomplished.

I was surprised and delighted when I explained the difficulties of the assignment to Mario that he readily agreed to put in his best effort to accomplish the task. Thus, beginning in December, Mario would spend several hours during his weekly visit researching out each of the several thousand chemicals on the internet and entering the required information into our database. To his credit, he stayed on this assignment till its completion today, April 2, 2010.

During this assignment, Mario was able to carry out the task with little supervision by me and only needed my assistance when he noticed some unexpected problem in the database or other errors.

Now that the work has been done, I have reviewed it and found the entries he made correct and the databases he used for reference to be highly regarded by myself and the biosafety committee members.

One of the most important characteristics that Mario brought to this assignment in addition to his willingness to work diligently on a very tedious and repetitive assignment was a very positive attitude towards the work itself. This was well received by his coworkers and myself.

In conclusion, I am very satisfied with the work that Mario Nabliba performed for Research Service and would recommend him for any other assignment.

Jerry Dungan
BioSafety Officer
Research Service

To: Whom it may concern

From: Rai Flowers – Author/Podcast Host (Book Of Secrets For Today's World) and Godis365 Podcast

Re: Mario Nabliba

Mario Nabliba is an exceptional individual. I have known him for over 30 years as a friend and former colleague. Under the United States Peace Corps umbrella, I served as an ESL English coordinator for Cape Verde, West Africa. I helped to coordinate English as a second language for Olavo Moniz HS on the island of Sal. Mario served as a math teacher and coordinator at the same high school. This is where we met and collaborated with other Cape Verdean teachers to help develop a curriculum that offers students the opportunity to receive a quality education. We worked closely with the Cape Verde Ministry of Education to ensure the curriculum served the goals and objectives desired by the Cape Verde government.

Mario was well liked and respected by his fellow colleagues and educational administrators. His leadership was evident in his communication and collaboration with parents and the community. It was no surprise to me to learn that Mario had become an established author after our service terminated. Over the years, Mario has proven himself to be a leader in his field, with his wisdom and integrity always shining through. His ability to attract positive friendships and his sense of humor are additional qualities that make him a joy to work with.

Without any doubts or hesitation, I would be more than willing to recommend and support Mario in any endeavor he wishes to pursue. His friendship and support have been invaluable to me, and I am confident he will bring the same dedication and excellence to any future project or position.

Sincerely,

Rai Flowers

www.ingramcontent.com/pod-product-compliance
Lightning Source LLC
Chambersburg PA
CBHW040908210326
41597CB00029B/5016